Season's Griddings

Graphing Activities for Listening
and
Following Directions

Grades 1–3

by
Will C. Howell

Published by Fearon Teacher Aids
an imprint of
Frank Schaffer Publications®

Author: Will C. Howell
Editors: Hanna Otero, Sarah Durand
Interior Artist: Corbin Hillam

Fearon Teacher Aids is an imprint of Frank Schaffer Publications.

Send all inquiries to:
Frank Schaffer Publications
3195 Wilson Drive NW
Grand Rapids, Michigan 49534

Season's Griddings—grades 1–3

ISBN: 0-7682-0800-9

3 4 5 6 7 8 9 10 SNC 10 09 08 07 06 05

Table of Contents

Introduction

The activities in *Seasons Griddings* are designed to develop listening skills, provide practice in following directions, reinforce left/right awareness, and provide readiness for coordinate geometry and graphing. Whether listening or reading, students will follow directions and move their pencils to create clever pictures that are fun to make. Before you begin, a bit of organizing information will be helpful.

Contents

Seasons Griddings has twenty-six reproducible holiday and seasonal activities. The activities are organized by month. They are designed for one-inch, one-half-inch or one-fourth-inch grid paper. Reproducible blackline masters of the grid papers are located in the back of the book.

Directions

Directions for the activities are presented in two distinct formats: descriptive directions and coordinate directions. For descriptive directions, a starting point is located and then students are directed to move along the lines of the grid paper. For example: "Move your pencil left 3 spaces, down 2 spaces, right 1 space, up 4 spaces . . . ."

Coordinate directions focus on a starting grid point and then direct the student to move from one grid point to another. For example: "Start at (–3,2). Move your pencil to (–3,5)." Coordinate directions are provided on separate reproducible activity sheets.

All activity directions may be read aloud to students, or duplicated and distributed for students to complete independently.

Answer Key

For each activity, a reduced image of the final product is provided on the same page as the descriptive directions. This reduced image can be used as the answer key for the activity. The images are the same for the final products of both descriptive and coordinate directions.

Preparation

Select an activity and duplicate the appropriate grid paper. Younger students should begin with one-inch grid activities. If you plan to distribute the activity sheets with descriptive directions, simply cover the answer-key image with a blank piece of paper before duplicating. Students will need a sheet of grid paper, a pencil, and an eraser. They will need an activity sheet if they are to be working independently.

Using Descriptive Directions

Activity directions indicate whether the paper is to be placed horizontally or vertically.

horizontally vertically

Next, students are directed to find the starting point by locating a beginning horizontal or vertical line and then counting a given number of spaces in a specific direction. Students should move their pencils along lines of the grid paper as they count the given number of full spaces. (Suggest that they think of the lines as streets and their pencils as cars. "Do not drive off the street into anyone's yard.") Full spaces are blocks that have all four corners visible. (Grid paper duplicated from the back of this book will not present a problem in locating the starting point or counting full spaces; however, commercial graph paper may have edges with spaces that are not complete. If you use commercially printed graph paper, be sure your students understand the concept of full spaces.)

Moving the Pencil

Students construct the picture by drawing a continuous line from the starting point; their pencils should not be lifted from their papers. A teaching suggestion for younger students: you might have them mark the left (L) and right (R) edges of their papers to help them avoid reversals.

start up1 space right 2 spaces up 2 spaces

Using Coordinate Directions

For the one-half-inch coordinate directions, students will locate the starting points by finding a letter or a number across the bottom and a number along the left edge of the grid. For example: A–1

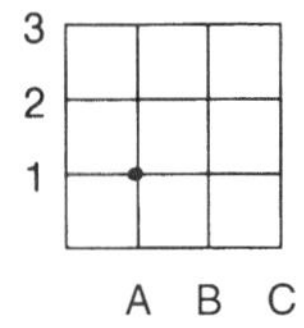

For one-half inch grid pictures, students should find the letter line and the number line and follow each line until the two intersect. The intersection of the horizontal and vertical lines is the starting point. Students locate each point in sequence and move their pencils continuously.

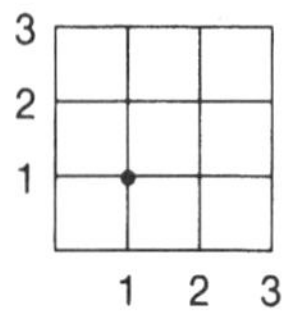

1–1

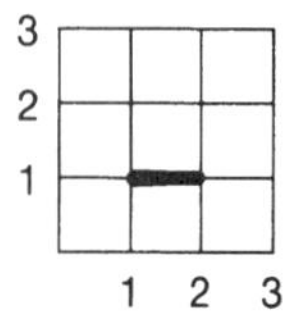

2–1

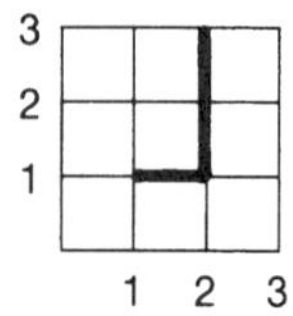

2–3

For one-fourth inch grid pictures, students should locate the horizontal and vertical number lines and follow each until the two intersect. The intersection of the horizontal and vertical lines is the starting point. As students locate each point in sequence, they move their pencils continuously.

Completing the Picture

After completing all the steps of an activity, students are directed to add details to their drawings, such as ears, eyes, and noses.

Quincy Quarterback

Paper: one-inch grid

Starting Point:

1. Place your paper with the short edge near the bottom of your desk.
2. Find the top end of the line closest to the left side of your paper.
3. Count down three full spaces.
4. Move right two full spaces.
5. Make a pencil dot where the lines cross.
6. Put your pencil on the dot.

Move your pencil:

1. Up 2 spaces
2. Right 2 spaces
3. Down 1 space
4. Left 1 space
5. Down 1 space
6. Left 3 spaces
7. Down 3 spaces
8. Right 1 space
9. Down 4 spaces
10. Right 2 spaces
11. Up 1 space
12. Left 1 space
13. Up 2 spaces
14. Right 5 spaces
15. Up 2 spaces
16. Left 1 space
17. Down 1 space
18. Left 2 spaces
19. Up 2 spaces
20. Right 3 spaces
21. Up 1 space
22. Left 3 spaces
23. Up 1 space

Add: eyes, nose, mouth, football, line for left arm, and number on his shirt

An Apple for My Teacher

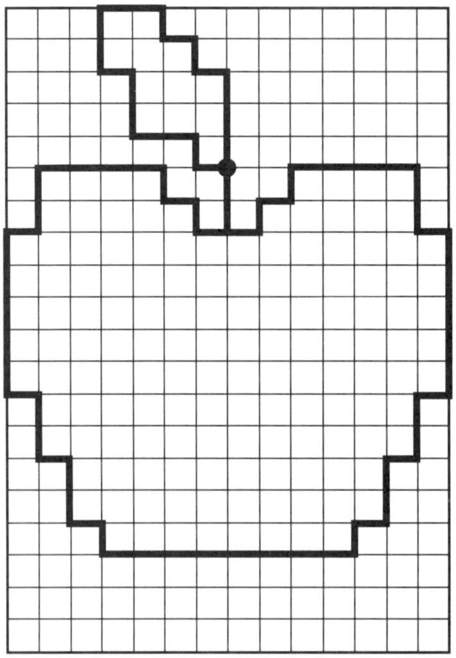

Paper: one-half-inch grid

Starting Point:

1. Place your paper with the short edge near the bottom of your desk.
2. Find the left end of the line closest to the top of your paper.
3. Count seven full spaces to the right.
4. Move down five full spaces.
5. Make a pencil dot where the lines cross.
6. Put your pencil on the dot.

Move your pencil:

1. Left 1 space
2. Up 1 space
3. Left 2 spaces
4. Up 2 spaces
5. Left 1 space
6. Up 2 spaces
7. Right 2 spaces
8. Down 1 space
9. Right 1 space
10. Down 1 space
11. Right 1 space
12. Down 5 spaces
13. Right 1 space
14. Up 1 space
15. Right 1 space
16. Up 1 space
17. Right 4 spaces
18. Down 2 spaces
19. Right 1 space
20. Down 5 spaces
21. Left 1 space
22. Down 2 spaces
23. Left 1 space
24. Down 2 spaces
25. Left 1 space
26. Down 1 space
27. Left 8 spaces
28. Up 1 space
29. Left 1 space
30. Up 2 spaces
31. Left 1 space
32. Up 2 spaces
33. Left 1 space
34. Up 5 spaces
35. Right 1 space
36. Up 2 spaces
37. Right 4 spaces
38. Down 1 space
39. Right 1 space
40. Down 1 space
41. Right 1 space

Johnny Appleseed

Paper: one-fourth-inch grid

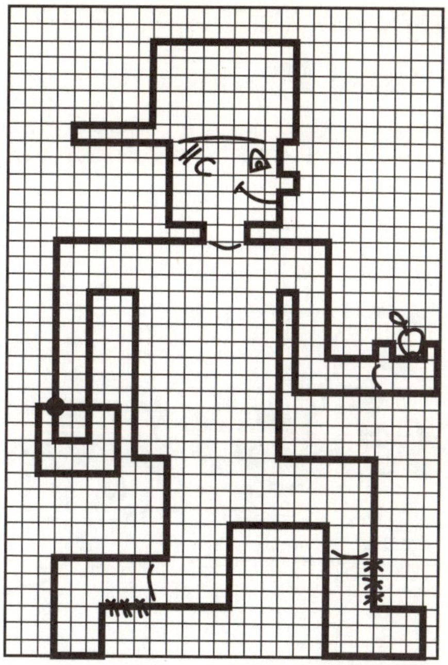

Starting Point:

1. Place your paper with the short edge near the bottom of your desk.
2. Find the left end of the line closest to the bottom of your paper.
3. Count up fifteen full spaces.
4. Move right three full spaces.
5. Make a pencil dot where the lines cross.
6. Put your pencil on the dot.

Move your pencil:

1. Left 1 space
2. Down 4 spaces
3. Right 5 spaces
4. Up 4 spaces
5. Left 2 spaces
6. Down 2 spaces
7. Left 2 spaces
8. Up 12 spaces
9. Right 9 spaces
10. Up 1 space
11. Left 2 spaces
12. Up 5 spaces
13. Left 6 spaces
14. Up 1 space
15. Right 5 spaces
16. Up 5 spaces
17. Right 9 spaces
18. Down 6 spaces
19. Left 1 space
20. Down 2 spaces
21. Right 1 space
22. Down 1 space
23. Left 1 space
24. Down 2 spaces
25. Left 2 spaces
26. Down 1 space
27. Right 5 spaces
28. Down 7 spaces
29. Right 3 spaces
30. Up 1 space
31. Right 1 space
32. Down 1 space
33. Right 2 spaces
34. Up 1 space
35. Right 1 space
36. Down 3 spaces
37. Left 9 spaces
38. Up 6 spaces
39. Left 1 space
40. Down 10 spaces
41. Right 6 spaces
42. Down 9 spaces
43. Right 3 spaces
44. Down 3 spaces
45. Left 6 spaces
46. Up 8 spaces
47. Left 6 spaces
48. Down 5 spaces
49. Left 8 spaces
50. Down 3 spaces
51. Left 3 spaces
52. Up 6 spaces
53. Right 7 spaces
54. Up 6 spaces
55. Left 2 spaces
56. Up 10 spaces
57. Left 3 spaces
58. Down 7 spaces
59. Left 2 spaces

Add: edge of pot, eye, smile, ear, hair, collar and cuffs, boots with laces, and apple

Johnny Appleseed

Paper: one-fourth-inch coordinate axis grid

Starting Point: Place your paper with the long edge near the bottom of your desk. Start at (–3, 11).

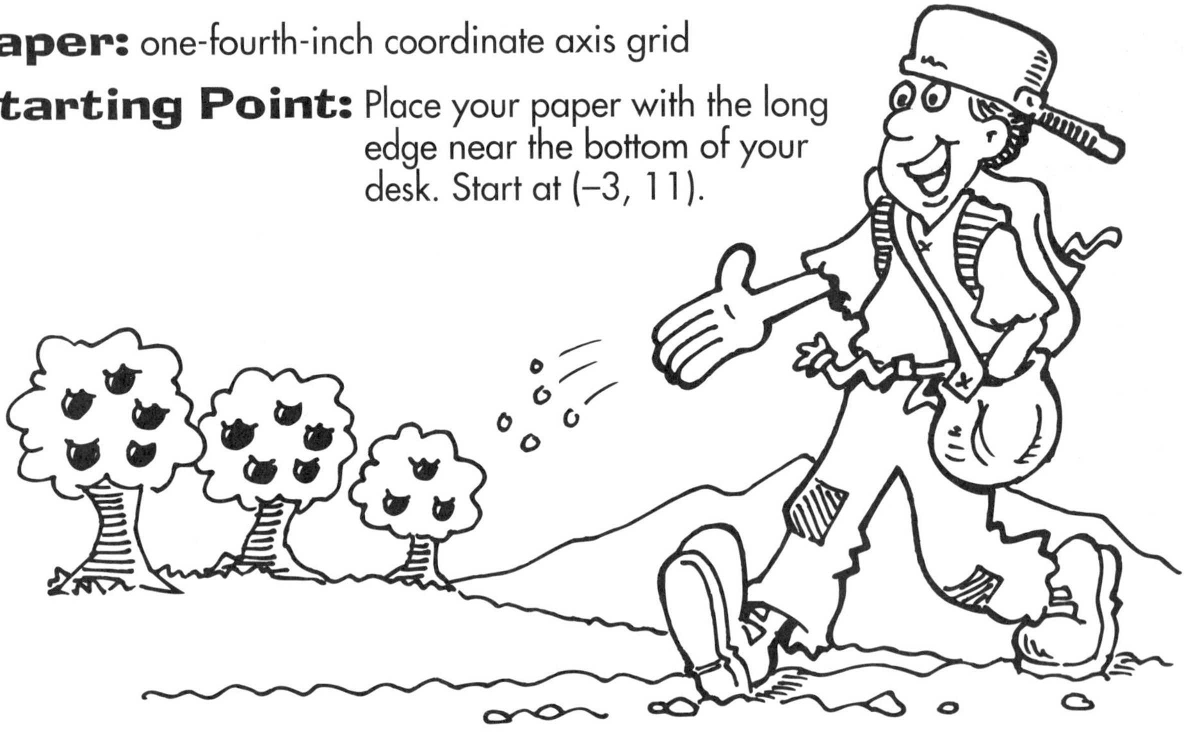

Move your pencil to:

1. (–3, 12)
2. (–7, 12)
3. (–7, 7)
4. (–3, 7)
5. (–3, 9)
6. (–5, 9)
7. (–5, 11)
8. (7, 11)
9. (7, 2)
10. (8, 2)
11. (8, 4)
12. (13, 4)
13. (13, 10)
14. (14, 10)
15. (14, 5)
16. (19, 5)
17. (19, –4)
18. (13, –4)
19. (13, –3)
20. (11, –3)
21. (11, –4)
22. (10, –4)
23. (10, –3)
24. (8, –3)
25. (8, –1)
26. (7, –1)
27. (7, –6)
28. (0, –6)
29. (0, –9)
30. (1, –9)
31. (1, –10)
32. (0, –10)
33. (0, –12)
34. (1, –12)
35. (1, –13)
36. (–2, –13)
37. (–2, –4)
38. (4, –4)
39. (4, –3)
40. (–6, –3)
41. (–6, –9)
42. (–15, –9)
43. (–15, –12)
44. (–18, –12)
45. (–18, –6)
46. (–10, –6)
47. (–10, 0)
48. (–15, 0)
49. (–15, 8)
50. (–18, 8)
51. (–18, 11)
52. (–12, 11)
53. (–12, 4)
54. (–6, 4)
55. (–6, 6)
56. (4, 6)
57. (4, 9)
58. (–3, 9)
59. (–3, 11)

Turn your paper and add: edge of pot, eye, smile, ear, hair, collar and cuffs, boots with laces, and apple

0-7682-0800-9 *Season's Griddings*

Trick or Treat Kitty

Paper: one-inch grid

Starting Point:

1. Place you paper with the short edge near the bottom of your desk.
2. Find the left end of the line closest to the bottom of your paper.
3. Count up two full spaces.
4. Move right two full spaces.
5. Make a pencil dot where the lines cross.
6. Put your pencil on the dot.

Move your pencil:

1. Right 1 space
2. Down 1 space
3. Right 1 space
4. Down 1 space
5. Right 2 spaces
6. Up 1 space
7. Left 1 space
8. Up 4 spaces
9. Right 1 space
10. Up 2 spaces
11. Left 1 space
12. Up 1 space
13. Left 1 space
14. Down 4 spaces
15. Left 2 spaces
16. Down 3 spaces
17. Left 1 space
18. Up 5 spaces
19. Right 1 space
20. Up 1 space
21. Left 2 spaces
22. Down 7 spaces
23. Right 4 spaces

Add: eye, nose, mouth, and whiskers

Golly Goal Post

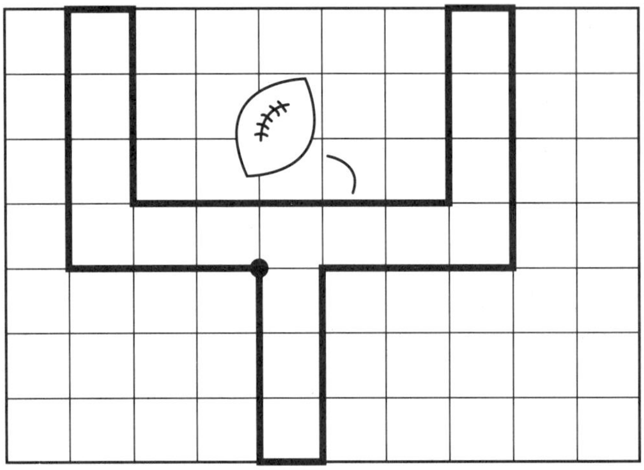

Paper: one-inch grid

Starting Point:

1. Place your paper with the long edge near the bottom of your desk.
2. Find the left end of the line closest to the bottom of your paper.
3. Count four full spaces to the right.
4. Count up three full spaces.
5. Make a pencil dot where the lines cross.
6. Put your pencil on the dot.

Move your pencil:

1. Left 3 spaces
2. Up 4 spaces
3. Right 1 space
4. Down 3 spaces
5. Right 5 spaces
6. Up 3 spaces
7. Right 1 space
8. Down 4 spaces
9. Left 3 spaces
10. Down 3 spaces
11. Left 1 space
12. Up 3 spaces

Add: a football between the posts

Scaredy Cat

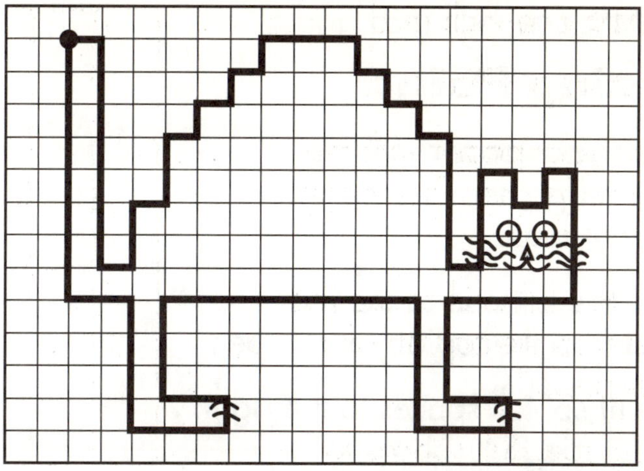

Paper: one-half-inch grid

Starting Point:

1. Place your paper with the long edge near the bottom of your desk.
2. Find the left end of the line closest to the top of your paper.
3. Count two full spaces to the right.
4. Move down one full space.
5. Make a pencil dot where the lines cross.
6. Put your pencil on the dot.

Move your pencil:

1. Right 1 space
2. Down 7 spaces
3. Right 1 space
4. Up 2 spaces
5. Right 1 space
6. Up 2 spaces
7. Right 1 space
8. Up 1 space
9. Right 1 space
10. Up 1 space
11. Right 1 space
12. Up 1 space
13. Right 3 spaces
14. Down 1 space
15. Right 1 space
16. Down 1 space
17. Right 1 space
18. Down 1 space
19. Right 1 space
20. Down 4 spaces
21. Right 1 space
22. Up 3 spaces
23. Right 1 space
24. Down 1 space
25. Right 1 space
26. Up 1 space
27. Right 1 space
28. Down 4 spaces
29. Left 4 spaces
30. Down 3 paces
31. Right 2 spaces
32. Down 1 space
33. Left 3 spaces
34. Up 4 spaces
35. Left 8 spaces
36. Down 3 spaces
37. Right 2 spaces
38. Down 1 space
39. Left 3 spaces
40. Up 4 spaces
41. Left 2 spaces
42. Up 8 spaces

Add: eyes, nose, mouth, whiskers, and claws

Rollicking Rake

Paper: one-half-inch grid

Starting Point:

1. Place your paper with the short edge near the bottom of your desk.
2. Find the left end of the line closest to the top of your paper.
3. Count six full spaces to the right.
4. Move your pencil down one full space.
5. Make a pencil dot where the lines cross.
6. Put your pencil on the dot.

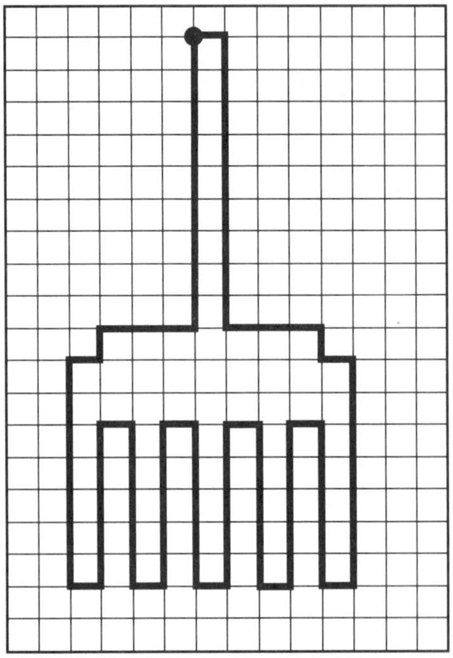

Move your pencil:

1. Down 9 spaces
2. Left 3 spaces
3. Down 1 space
4. Left 1 space
5. Down 7 spaces
6. Right 1 space
7. Up 5 spaces
8. Right 1 space
9. Down 5 spaces
10. Right 1 space
11. Up 5 spaces
12. Right 1 space
13. Down 5 spaces
14. Right 1 space
15. Up 5 spaces
16. Right 1 space
17. Down 5 spaces
18. Right 1 space
19. Up 5 spaces
20. Right 1 space
21. Down 5 spaces
22. Right 1 space
23. Up 7 spaces
24. Left 1 space
25. Up 1 space
26. Left 3 spaces
27. Up 9 spaces
28. Left 1 space

Scaredy Cat

Paper: one-half-inch letter coordinate grid

Starting Point: Place your paper with the long edge near the bottom of your desk. Start at B–13.

Move your pencil to:

1. C–13
2. C–6
3. D–6
4. D–8
5. E–8
6. E–10
7. F–10
8. F–11
9. G–11
10. G–12
11. H–12
12. H–13
13. K–13
14. K–12
15. L–12
16. L–11
17. M–11
18. M–10
19. N–10
20. N–6
21. O–6
22. O–9
23. P–9
24. P–8
25. Q–8
26. Q–9
27. R–9
28. R–5
29. N–5
30. N–2
31. P–2
32. P–1
33. M–1
34. M–5
35. E–5
36. E–2
37. G–2
38. G–1
39. D–1
40. D–5
41. B–5
42. B–13

Add: eyes, nose, mouth, whiskers, and claws

Rollicking Rake

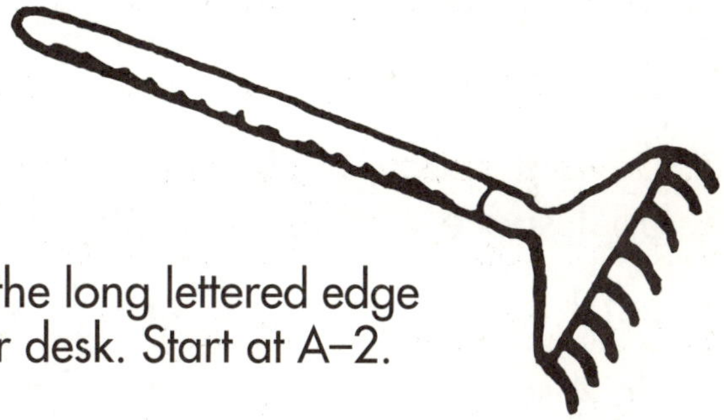

Paper: one-half-inch letter coordinate grid

Starting Point: Place your paper with the long lettered edge near the bottom of your desk. Start at A–2.

Move your pencil to:

1. A–3
2. F–3
3. F–4
4. A–4
5. A–5
6. F–5
7. F–6
8. A–6
9. A–7
10. F–7
11. F–8
12. A–8
13. A–9
14. F–9
15. F–10
16. A–10
17. A–11
18. H–11
19. H–10
20. I–10
21. I–7
22. R–7
23. R–6
24. I–6
25. I–3
26. H–3
27. H–2
28. A–2
29. Turn your paper so the short numbered edge is near the bottom of your desk.

Perfect Pumpkin

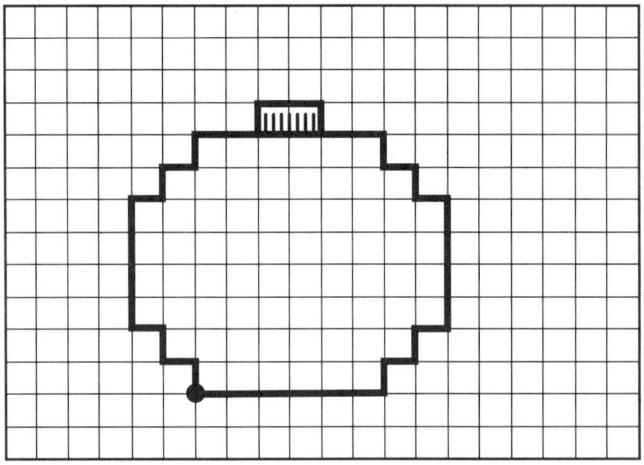

Paper: one-half-inch grid

Starting Point:

1. Place your paper with the long edge near the bottom of your desk.
2. Find the left end of the line closest to the bottom of your paper.
3. Count six full spaces to the right.
4. Move up two full spaces.
5. Make a pencil dot where the lines cross.
6. Put your pencil on the dot.

Move your pencil:

1. Up 1 space
2. Left 1 space
3. Up 1 space
4. Left 1 space
5. Up 4 spaces
6. Right 1 space
7. Up 1 space
8. Right 1 space
9. Up 1 space
10. Right 4 spaces
11. Up 1 space
12. Left 2 spaces
13. Down 1 space
14. Right 4 spaces (return over your line)
15. Down 1 space
16. Right 1 space
17. Down 1 space
18. Right 1 space
19. Down 4 spaces
20. Left 1 space
21. Down 1 space
22. Left 1 space
23. Down 1 space
24. Left 6 spaces

Add: lines on stem

Lofty Leaf

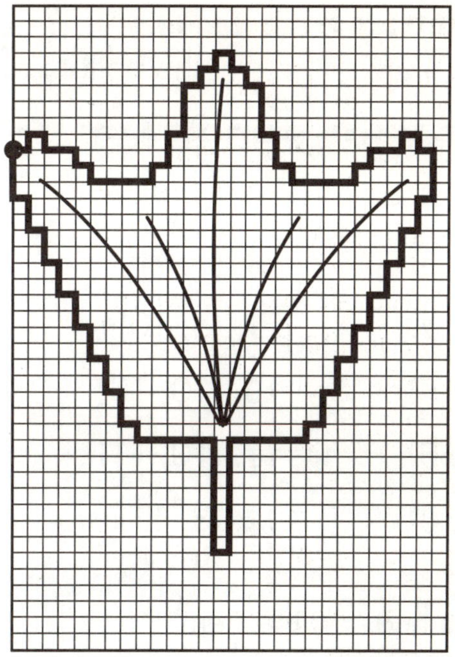

Paper: one-fourth-inch grid

Starting Point:

1. Place your paper with the short edge near the bottom of your desk.
2. Find the left end of the line closest to the top of your paper.
3. Count down nine full spaces.
4. Make a pencil dot where the lines cross.
5. Put your pencil on the dot.

Move your pencil:

1. Right 1 space
2. Up 1 space
3. Right 1 space
4. Down 1 space
5. Right 2 spaces
6. Down 1 space
7. Right 1 space
8. Down 1 space
9. Right 4 spaces
10. Up 1 space
11. Right 1 space
12. Up 2 spaces
13. Right 1 space
14. Up 3 spaces
15. Right 1 space
16. Up 1 space
17. Right 1 space
18. Up 1 space
19. Right 1 space
20. Down 1 space
21. Right 1 space
22. Down 1 space
23. Right 1 space
24. Down 3 spaces
25. Right 1 space
26. Down 2 spaces
27. Right 1 space
28. Down 1 space
29. Right 4 spaces
30. Up 1 space
31. Right 1 space
32. Up 1 space
33. Right 2 spaces
34. Up 1 space
35. Right 1 space
36. Down 1 space
37. Right 1 space
38. Down 3 spaces
39. Left 1 space
40. Down 2 spaces
41. Left 1 space
42. Down 2 spaces
43. Left 1 space
44. Down 2 spaces
45. Left 1 space
46. Down 2 spaces
47. Left 1 space
48. Down 2 spaces
49. Left 1 space
50. Down 2 spaces
51. Left 1 space
52. Down 2 spaces
53. Left 1 space
54. Down 1 space
55. Left 5 spaces
56. Down 7 spaces
57. Left 1 space
58. Up 7 spaces
59. Left 5 spaces
60. Up 1 space
61. Left 1 space
62. Up 2 spaces
63. Left 1 space
64. Up 2 spaces
65. Left 1 space
66. Up 2 spaces
67. Left 1 space
68. Up 2 spaces
69. Left 1 space
70. Up 2 spaces
71. Left 1 space
72. Up 2 spaces
73. Left 1 space
74. Up 2 spaces
75. Left 1 space
76. Up 3 spaces

Add: veins on the leaf

0-7682-0800-9 *Season's Griddings*

Perfect Pumpkin

Paper: one-half-inch number coordinate grid

Starting Point: Place your paper with the long edge near the bottom of your desk. Start at (6, 2).

Move your pencil to:

1. (6, 3)	**7.** (5, 9)	**13.** (8, 10)	**19.** (14, 4)
2. (5, 3)	**8.** (6, 9)	**14.** (12, 10)	**20.** (13, 4)
3. (5, 4)	**9.** (6, 10)	**15.** (12, 9)	**21.** (13, 3)
4. (4, 4)	**10.** (10, 10)	**16.** (13, 9)	**22.** (12, 3)
5. (4, 8)	**11.** (10, 11)	**17.** (13, 8)	**23.** (12, 2)
6. (5, 8)	**12.** (8, 11)	**18.** (14, 8)	**24.** (6, 2)

Add: lines on the stem

Lofty Leaf

Paper: one-fourth-inch coordinate axis grid

Starting Point: Place your paper with the long edge near the bottom of your desk. Start at (–12, 0).

Move your pencil to:

1. (–12, 1)	**14.** (6, 11)	**27.** (12, 9)	**40.** (18, 0)	**53.** (13, –11)	**66.** (2, –9)
2. (–5, 1)	**15.** (6, 12)	**28.** (11, 9)	**41.** (18, –1)	**54.** (14, –11)	**67.** (2, –8)
3. (–5, 6)	**16.** (8, 12)	**29.** (11, 5)	**42.** (17, –1)	**55.** (14, –12)	**68.** (0, –8)
4. (–4, 6)	**17.** (8, 13)	**30.** (12, 5)	**43.** (17, –2)	**56.** (13, –12)	**69.** (0, –7)
5. (–4, 7)	**18.** (10, 13)	**31.** (12, 4)	**44.** (14, –2)	**57.** (13, –13)	**70.** (–2, –7)
6. (–2, 7)	**19.** (10, 14)	**32.** (14, 4)	**45.** (14, –3)	**58.** (10, –13)	**71.** (–2, –6)
7. (–2, 8)	**20.** (13, 14)	**33.** (14, 3)	**46.** (12, –3)	**59.** (10, –12)	**72.** (–4, –6)
8. (0, 8)	**21.** (13, 13)	**34.** (17, 3)	**47.** (12, –4)	**60.** (8, –12)	**73.** (–4, –5)
9. (0, 9)	**22.** (14, 13)	**35.** (17, 2)	**48.** (11, –4)	**61.** (8, –11)	**74.** (–5, –5)
10. (2, 9)	**23.** (14, 12)	**36.** (18, 2)	**49.** (11, –8)	**62.** (6, –11)	**75.** (–5, 0)
11. (2, 10)	**24.** (13, 12)	**37.** (18, 1)	**50.** (12, –8)	**63.** (6, –10)	**76.** (–12, 0)
12. (4, 10)	**25.** (13, 10)	**38.** (19, 1)	**51.** (12, –9)	**64.** (4, –10)	
13. (4, 11)	**26.** (12, 10)	**39.** (19, 0)	**52.** (13, –9)	**65.** (4, –9)	

Turn your paper and add: veins on the leaf

Super Sailing

Paper: one-fourth-inch grid

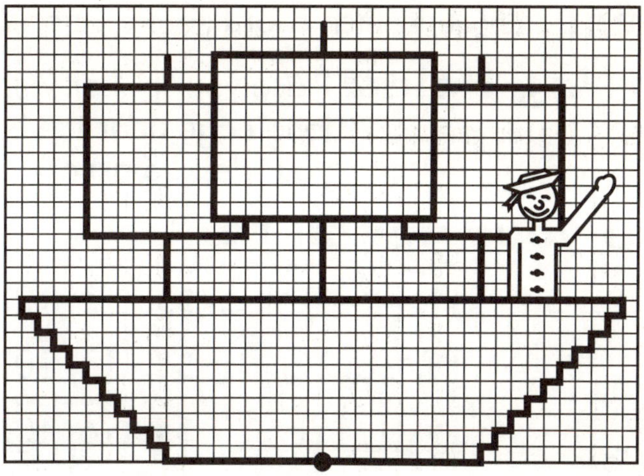

Starting Point:

1. Place your paper with the long edge near the bottom of your desk.
2. Find the left end of the line closest to the bottom of your paper.
3. Move 20 spaces to the right.
4. Make a pencil dot where the lines cross.
5. Put your pencil on the dot.

Move your pencil:

1. Left 10 spaces
2. Up 1 space
3. Left 1 space
4. Up 1 space
5. Left 1 space
6. Up 1 space
7. Left 1 space
8. Up 1 space
9. Left 1 space
10. Up 1 space
11. Left 1 space
12. Up 1 space
13. Left 1 space
14. Up 1 space
15. Left 1 space
16. Up 1 space
17. Left 1 space
18. Up 1 space
19. Left 1 space
20. Up 1 space
21. Right 9 spaces
22. Up 4 spaces
23. Left 5 spaces
24. Up 9 spaces
25. Right 5 spaces
26. Up 2 spaces
27. Down 2 spaces (return over your line)
28. Right 3 spaces
29. Down 8 spaces
30. Right 2 spaces
31. Down 1 space
32. Left 5 spaces
33. Down 4 spaces (return over your line)
34. Right 10 spaces
35. Up 5 spaces
36. Left 7 spaces (return over your line)
37. Up 10 spaces (return over your line)
38. Right 7 spaces
39. Up 2 spaces
40. Down 2 spaces (return over your line)
41. Right 7 spaces
42. Down 10 spaces
43. Left 7 spaces
44. Down 5 spaces (return over your line)
45. Right 10 spaces
46. Up 4 spaces
47. Left 5 spaces
48. Up 1 space
49. Right 2 spaces (return over your line)
50. Up 8 spaces (return over your line)
51. Right 3 spaces
52. Up 2 spaces
53. Down 2 spaces (return over your line)
54. Right 5 spaces
55. Down 9 spaces
56. Left 5 spaces
57. Down 4 spaces (return over your line)
58. Right 9 spaces
59. Down 1 space
60. Left 1 space
61. Down 1 space
62. Left 1 space
63. Down 1 space
64. Left 1 space
65. Down 1 space
66. Left 1 space
67. Down 1 space
68. Left 1 space
69. Down 1 space
70. Left 1 space
71. Down 1 space
72. Left 1 space
73. Down 1 space
74. Left 1 space
75. Down 1 space
76. Left 1 space
77. Down 1 space
78. Left 10 spaces

Add: a sailor on the deck

Super Saililng

Paper: one-fourth-inch coordinate axis grid

Starting Point: Place your paper with the long edge near the bottom of your desk. Start at (–18, –4).

Move your pencil to:

1. (–18, –3)
2. (–9, –3)
3. (–9, 1)
4. (–14, 1)
5. (–14, 10)
6. (–9, 10)
7. (–9, 12)
8. (–9, 10)
9. (–6, 10)
10. (–6, 2)
11. (–4, 2)
12. (–4, 1)
13. (–9, 1)
14. (–9, –3)
15. (1, –3)
16. (1, 2)
17. (–6, 2)
18. (–6, 12)
19. (1, 12)
20. (1, 14)
21. (1, 12)
22. (8, 12)
23. (8, 2)
24. (1, 2)
25. (1, –3)
26. (11, –3)
27. (11, 1)
28. (6, 1)
29. (6, 2)
30. (8, 2)
31. (8, 10)
32. (11, 10)
33. (11, 12)
34. (11, 10)
35. (16, 10)
36. (16, 1)
37. (11, 1)
38. (11, –3)
39. (21, –3)
40. (21, –4)
41. (20, –4)
42. (20, –5)
43. (19, –5)
44. (19, –6)
45. (18, –6)
46. (18, –7)
47. (17, –7)
48. (17, –8)
49. (16, –8)
50. (16, –9)
51. (15, –9)
52. (15, –10)
53. (14, –10)
54. (14, –11)
55. (13, –11)
56. (13, –12)
57. (12, –12)
58. (12, –13)
59. (–9, –13)
60. (–9, –12)
61. (–10, –12)
62. (–10, –11)
63. (–11, –11)
64. (–11, –10)
65. (–12, –10)
66. (–12, –9)
67. (–13, –9)
68. (–13, –8)
69. (–14, –8)
70. (–14, –7)
71. (–15, –7)
72. (–15, –6)
73. (–16, –6)
74. (–16, –5)
75. (–17, –5)
76. (–17, –4)
77. (–18, –4)

Add: a sailor on the deck

Everybody Vote!

Paper: one-inch grid (two pieces)

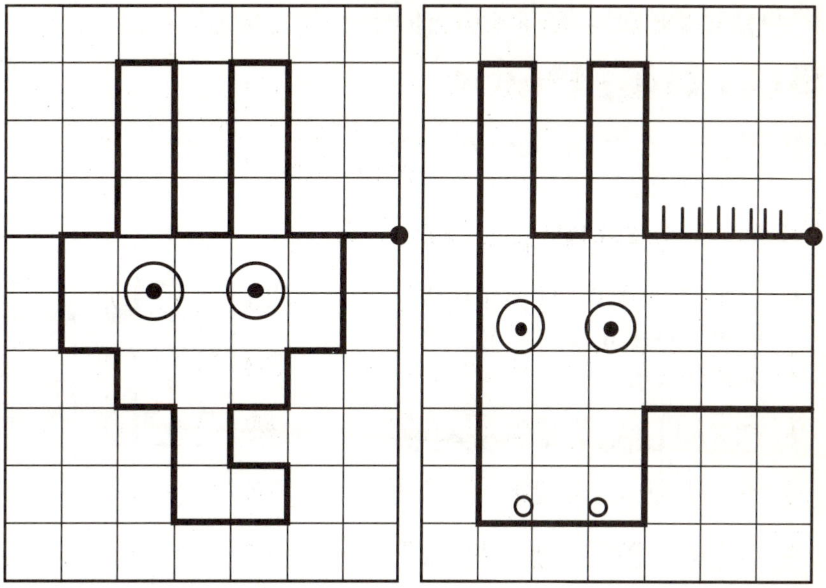

Starting Point:

1. Place paper with the short edge near the bottom of your desk.
2. Find the top end of the line closest to the right side of your paper.
3. Count down four full spaces.
4. Make a pencil dot where the lines cross.
5. Put your pencil on the dot.

Move your pencil:

(Republican)

1. Left 2 spaces
2. Up 3 spaces
3. Left 1 space
4. Down 3 spaces
5. Left 1 space
6. Up 3 spaces
7. Left 1 space
8. Down 3 spaces
9. Left 1 space
10. Down 2 spaces
11. Right 1 space
12. Down 1 space
13. Right 1 space
14. Down 2 spaces
15. Right 2 spaces
16. Up 1 space
17. Left 1 space
18. Up 1 space
19. Right 1 space
20. Up 1 space
21. Right 1 space
22. Up 2 spaces

Add: eyes

(Democrat)

1. Left 3 spaces
2. Up 3 spaces
3. Left 1 space
4. Down 3 spaces
5. Left 1 space
6. Up 3 spaces
7. Left 1 space
8. Down 8 spaces
9. Right 3 spaces
10. Up 2 spaces
11. Right 3 spaces

Add: eyes, nostrils and lines for hair

Mayflower

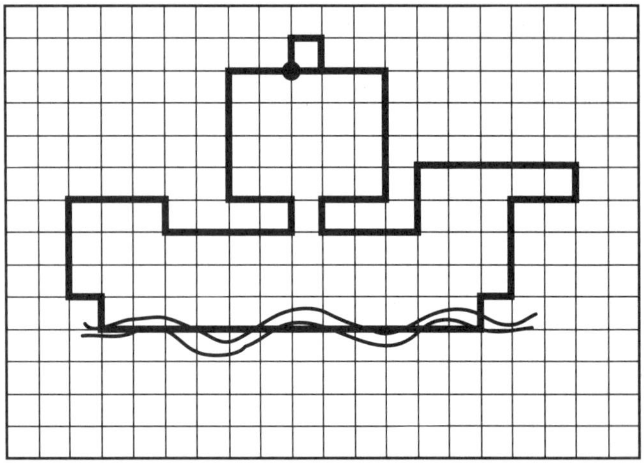

Paper: one-half-inch grid

Starting Point:

1. Place your paper with the long edge near the bottom of your desk.
2. Find the left end of the line closest to the top of your paper.
3. Count nine full spaces to the right.
4. Move down two full spaces.
5. Make a pencil dot where the lines cross.
6. Put your pencil on the dot.

Move your pencil:

1. Right 1 space
2. Up 1 space
3. Left 1 space
4. Down 1 space
5. Left 2 spaces
6. Down 4 spaces
7. Right 2 spaces
8. Down 1 space
9. Left 4 spaces
10. Up 1 space
11. Left 3 spaces
12. Down 3 spaces
13. Right 1 space
14. Down 1 space
15. Right 12 spaces
16. Up 1 space
17. Right 1 space
18. Up 3 spaces
19. Right 2 spaces
20. Up 1 space
21. Left 5 spaces
22. Down 2 spaces
23. Left 3 spaces
24. Up 1 space
25. Right 2 spaces
26. Up 4 spaces
27. Left 2 spaces

Add: waves

Timothy Turkey

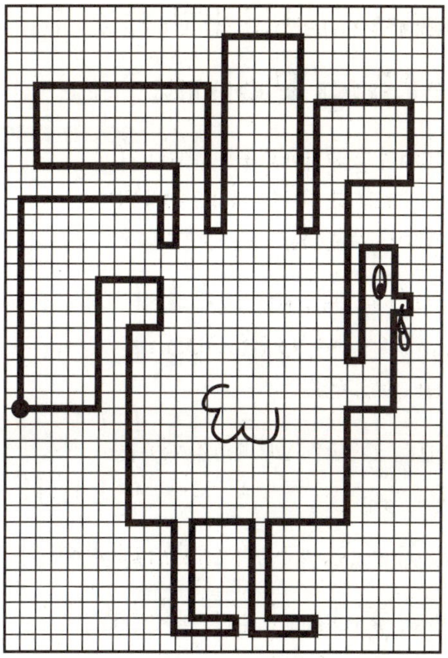

Paper: one-fourth-inch grid

Starting Point:

1. Place your paper with the short edge near the bottom of your desk.
2. Find the left end of the line closest to the bottom of your paper.
3. Count one full space to the right.
4. Move up fifteen full spaces.
5. Make a pencil dot where the lines cross.
6. Put your pencil on the dot.

Move your pencil:

1. Up 13 spaces
2. Right 9 spaces
3. Down 3 spaces
4. Right 1 space
5. Up 5 spaces
6. Left 9 spaces
7. Up 5 spaces
8. Right 11 spaces
9. Down 9 spaces
10. Right 1 space
11. Up 12 spaces
12. Right 5 spaces
13. Down 12 spaces
14. Right 1 space
15. Up 8 spaces
16. Right 6 spaces
17. Down 5 spaces
18. Left 4 spaces
19. Down 11 spaces
20. Right 1 space
21. Up 7 spaces
22. Right 2 spaces
23. Down 3 spaces
24. Right 1 space
25. Down 1 space
26. Left 1 space
27. Down 6 spaces
28. Left 3 spaces
29. Down 7 spaces
30. Left 5 spaces
31. Down 6 spaces
32. Right 3 spaces
33. Down 1 space
34. Left 4 spaces
35. Up 7 spaces
36. Left 4 spaces
37. Down 6 spaces
38. Right 3 spaces
39. Down 1 space
40. Left 4 spaces
41. Up 7 spaces
42. Left 3 spaces
43. Up 12 spaces
44. Right 2 spaces
45. Up 3 spaces
46. Left 4 spaces
47. Down 8 spaces
48. Left 5 spaces

Add: eye, wattle, and wing

Timothy Turkey

Paper: one-fourth-inch coordinate axis grid

Starting Point: Place your paper with the long edge near the bottom of your desk. Start at (5, –12).

Move your pencil to:

1. (–8, –12)
2. (–8, –3)
3. (–5, –3)
4. (–5, –2)
5. (–10, –2)
6. (–10, –11)
7. (–15, –11)
8. (–15, 0)
9. (–6, 0)
10. (–6, 1)
11. (–18, 1)
12. (–18, 6)
13. (–6, 6)
14. (–6, 7)
15. (–14, 7)
16. (–14, 13)
17. (–9, 13)
18. (–9, 9)
19. (2, 9)
20. (2, 10)
21. (–5, 10)
22. (–5, 12)
23. (–2,12)
24. (–2, 13)
25. (–1, 13)
26. (–1, 12)
27. (5, 12)
28. (5, 9)
29. (12, 9)
30. (12, 4)
31. (18, 4)
32. (18, 7)
33. (19, 7)
34. (19, 3)
35. (12, 3)
36. (12, –1)
37. (18, –1)
38. (18, 2)
39. (19, 2)
40. (19, –2)
41. (12, –2)
42. (12, –5)
43. (0, –5)
44. (0, –3)
45. (–3, –3)
46. (–3, –7)
47. (5, –7)
48. (5, –12)

Turn your paper and add: eye, wattle, and wing

Party Piñata

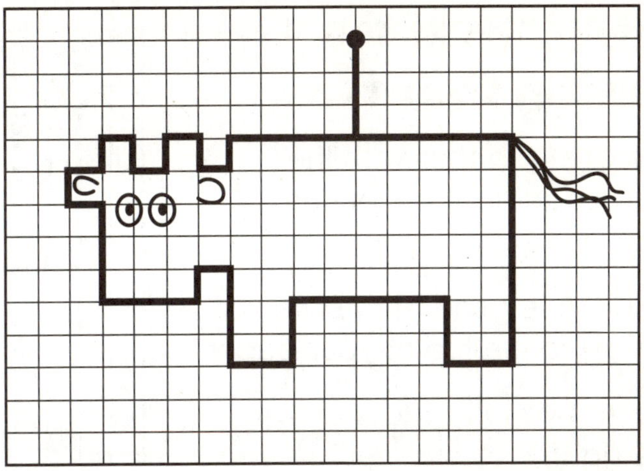

Paper: one-half-inch grid

Starting Point:

1. Place your paper with the long edge near the bottom of your desk.
2. Find the right end of the line closest to the top of your paper.
3. Count nine full spaces to the left.
4. Move down one full space.
5. Make a pencil dot where the lines cross.
6. Put your pencil on the dot.

Move your pencil:

1. Down 3 spaces
2. Left 4 spaces
3. Down 1 space
4. Left 1 space
5. Up 1 space
6. Left 1 space
7. Down 1 space
8. Left 1 space
9. Up 1 space
10. Left 1 space
11. Down 1 space
12. Left 1 space
13. Down 1 space
14. Right 1 space
15. Down 3 spaces
16. Right 3 spaces
17. Up 1 space
18. Right 1 space
19. Down 3 spaces
20. Right 2 spaces
21. Up 2 spaces
22. Right 5 spaces
23. Down 2 spaces
24. Right 2 spaces
25. Up 7 spaces
26. Left 5 spaces

Add: eyes, ears, and tail

Happy Hanukkah

Paper: one-half-inch grid

Starting Point:

1. Place your paper with the long edge near the bottom of your desk.
2. Find the left end of the line closest to the top of your paper.
3. Count two full spaces to the right.
4. Move down 3 full spaces.
5. Make a pencil dot where the lines cross.
6. Put your pencil on the dot.

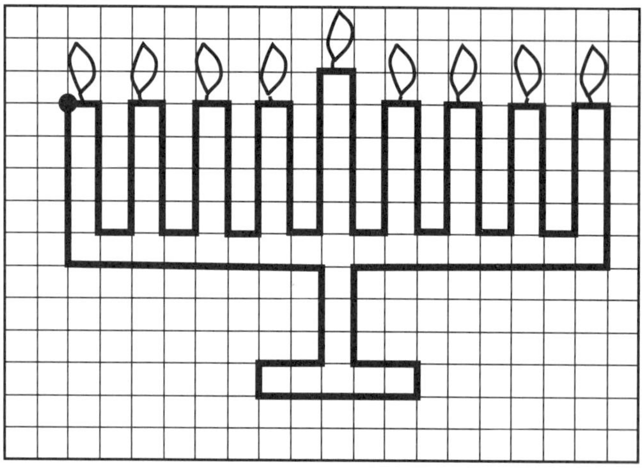

Move your pencil:

1. Right 1 space
2. Down 4 spaces
3. Right 1 space
4. Up 4 spaces
5. Right 1 space
6. Down 4 spaces
7. Right 1 space
8. Up 4 spaces
9. Right 1 space
10. Down 4 spaces
11. Right 1 space
12. Up 4 spaces
13. Right 1 space
14. Down 4 spaces
15. Right 1 space
16. Up 5 spaces
17. Right 1 space
18. Down 5 spaces
19. Right 1 space
20. Up 4 spaces
21. Right 1 space
22. Down 4 spaces
23. Right 1 space
24. Up 4 spaces
25. Right 1 space
26. Down 4 spaces
27. Right 1 space
28. Up 4 spaces
29. Right 1 space
30. Down 4 spaces
31. Right 1 space
32. Up 4 spaces
33. Right 1 space
34. Down 5 spaces
35. Left 8 spaces
36. Down 3 spaces
37. Right 2 spaces
38. Down 1 space
39. Left 5 spaces
40. Up 1 space
41. Right 2 spaces
42. Up 3 spaces
43. Left 8 spaces
44. Up 5 spaces

Add: flames

Party Piñata

Paper: one-half-inch letter coordinate grid

Starting Point: Place your paper with the long lettered edge near the bottom of your desk. Start at K–12.

Move your pencil to:

1. K–9
2. G–9
3. G–8
4. F–8
5. F–9
6. E–9
7. E–8
8. D–8
9. D–9
10. C–9
11. C–8
12. B–8
13. B–7
14. C–7
15. C–4
16. F–4
17. F–5
18. G–5
19. G–2
20. I–2
21. I–4
22. N–4
23. N–2
24. P–2
25. P–9
26. K–9

Add: eyes, ears, and tail

Happy Hanukkah

Paper: one-half-inch letter coordinate grid

Starting Point: Place your paper with the long lettered edge near the bottom of your desk. Start at A–11.

Move your pencil to:

1. B–11
2. B–7
3. C–7
4. C–11
5. D–11
6. D–7
7. E–7
8. E–11
9. F–11
10. F–7
11. G–7
12. G–11
13. H–11
14. H–7
15. I–7
16. I–12
17. J–12
18. J–7
19. K–7
20. K–11
21. L–11
22. L–7
23. M–7
24. M–11
25. N–11
26. N–7
27. O–7
28. O–11
29. P–11
30. P–7
31. Q–7
32. Q–11
33. R–11
34. R–6
35. J–6
36. J–3
37. L–3
38. L–2
39. G–2
40. G–3
41. I–3
42. I–6
43. A–6
44. A–11

Add: flames

Mr. Snow

Paper: one-half-inch grid

Starting Point:

1. Place your paper with the short edge near the bottom of your desk.
2. Find the right end of the line closest to the top of your paper.
3. Count four full spaces to the left.
4. Move down nine full spaces.
5. Make a pencil dot where the lines cross.
6. Put your pencil on the dot.

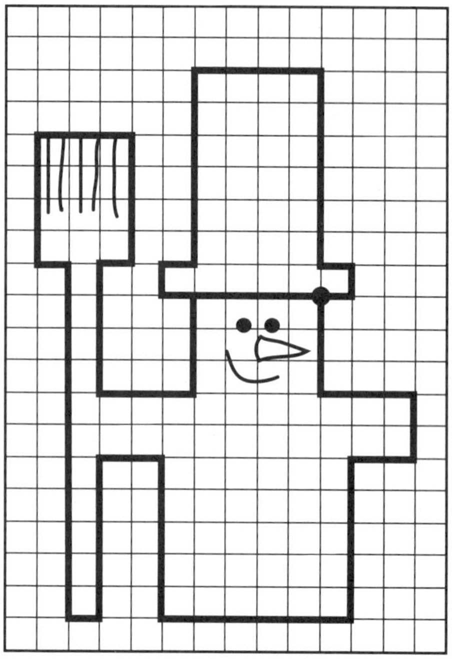

Move your pencil:

1. Left 5 spaces
2. Up 1 space
3. Right 1 space
4. Up 6 spaces
5. Right 4 spaces
6. Down 6 spaces
7. Right 1 space
8. Down 1 space
9. Left 1 space
10. Down 3 spaces
11. Right 3 spaces
12. Down 2 spaces
13. Left 2 spaces
14. Down 5 spaces
15. Left 6 spaces
16. Up 5 spaces
17. Left 2 spaces
18. Down 5 spaces
19. Left 1 space
20. Up 11 spaces
21. Left 1 space
22. Up 4 spaces
23. Right 3 spaces
24. Down 4 spaces
25. Left 1 space
26. Down 4 spaces
27. Right 3 spaces
28. Up 3 spaces

Add: eyes, nose, mouth, and bristles for broom

Randy Reindeer

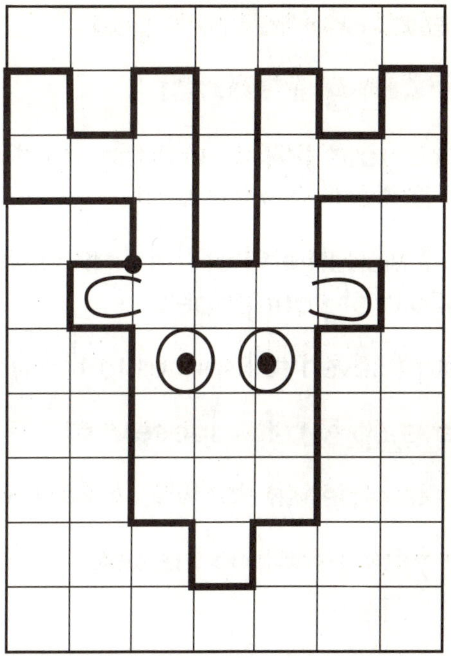

Paper: one-inch grid

Starting Point:

1. Place your paper with the short edge near the bottom of your desk.
2. Find the left end of the line closest to the top of your paper.
3. Count two full spaces to the right.
4. Move down four full spaces.
5. Make a pencil dot where the lines cross.
6. Put your pencil on the dot.

Move your pencil:

1. Up 1 space
2. Left 2 spaces
3. Up 2 spaces
4. Right 1 space
5. Down 1 space
6. Right 1 space
7. Up 1 space
8. Right 1 space
9. Down 3 spaces
10. Right 1 space
11. Up 3 spaces
12. Right 1 space
13. Down 1 space
14. Right 1 space
15. Up 1 space
16. Right 1 space
17. Down 2 spaces
18. Left 2 spaces
19. Down 1 space
20. Right 1 space
21. Down 1 space
22. Left 1 space
23. Down 3 spaces
24. Left 1 space
25. Down 1 space
26. Left 1 space
27. Up 1 space
28. Left 1 space
29. Up 3 spaces
30. Left 1 space
31. Up 1 space
32. Right 1 space

Add: eyes and ears

Prickly Pine

Paper: one-half-inch grid

Starting Point:

1. Place your paper with the short edge near the bottom of your desk.
2. Find the left end of the line closest to the bottom of your paper.
3. Count seven full spaces to the right.
4. Move up two full spaces.
5. Make a pencil dot where the lines cross.
6. Put your pencil on the dot.

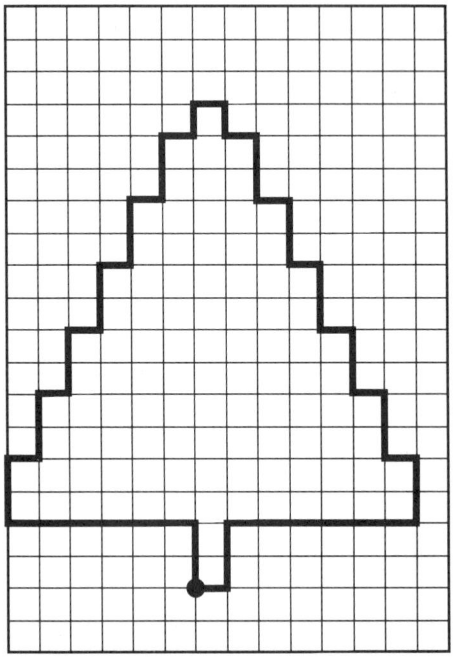

Move your pencil:

1. Up 2 spaces
2. Right 6 spaces
3. Up 2 spaces
4. Left 1 space
5. Up 2 spaces
6. Left 1 space
7. Up 2 spaces
8. Left 1 space
9. Up 2 spaces
10. Left 1 space
11. Up 2 spaces
12. Left 1 space
13. Up 2 spaces
14. Left 1 space
15. Up 1 space
16. Left 1 space
17. Down 1 space
18. Left 1 space
19. Down 2 spaces
20. Left 1 space
21. Down 2 spaces
22. Left 1 space
23. Down 2 spaces
24. Left 1 space
25. Down 2 spaces
26. Left 1 space
27. Down 2 spaces
28. Left 1 space
29. Down 2 spaces
30. Right 6 spaces
31. Down 2 spaces
32. Right 1 space

Prickly Pine

Paper: one-half-inch number coordinate grid

Starting Point: Place your paper with the long edge near the bottom of your desk. Start at (2, 6).

Move your pencil to:

1. (2, 7)
2. (4, 7)
3. (4, 13)
4. (6, 13)
5. (6, 12)
6. (8, 12)
7. (8, 11)
8. (10, 11)
9. (10, 10)
10. (12, 10)
11. (12, 9)
12. (14, 9)
13. (14, 8)
14. (16, 8)
15. (16, 7)
16. (17, 7)
17. (17, 6)
18. (16, 6)
19. 16, 5)
20. (14, 5)
21. (14, 4)
22. (12, 4)
23. (12, 3)
24. (10, 3)
25. (10, 2)
26. (8, 2)
27. (8, 1)
28. (6, 1)
29. (6, 0)
30. (4, 0)
31. (4, 6)
32. (2, 6)

Slick Slim's Sled

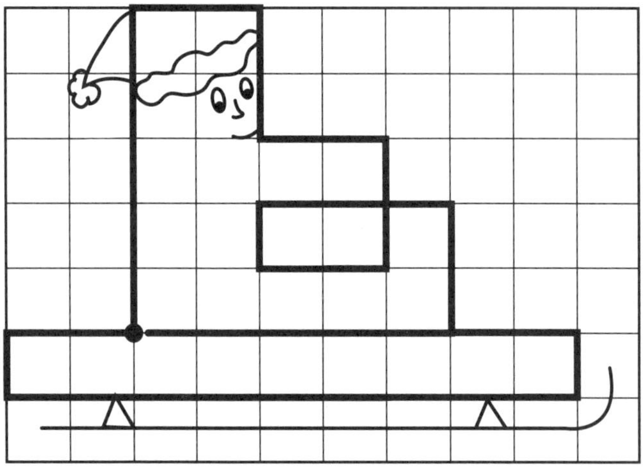

Paper: one-inch grid

Starting Point:

1. Place your paper with the long edge near the bottom of your desk.
2. Find the left end of the line closest to the bottom of your paper.
3. Count two full spaces to the right.
4. Move up two full spaces.
5. Make a pencil dot where the lines cross.
6. Put your pencil on the dot.

Move your pencil:

1. Right 5 spaces
2. Up 2 spaces
3. Left 1 space
4. Down 1 space
5. Left 2 spaces
6. Up 1 space
7. Right 2 spaces
8. Up 1 space
9. Left 2 spaces
10. Up 2 spaces
11. Left 2 spaces
12. Down 5 spaces
13. Left 2 spaces
14. Down 1 space
15. Right 9 spaces
16. Up 1 space
17. Left 2 spaces

Add: face, hat, and runner for the sled

Chinese Dragon

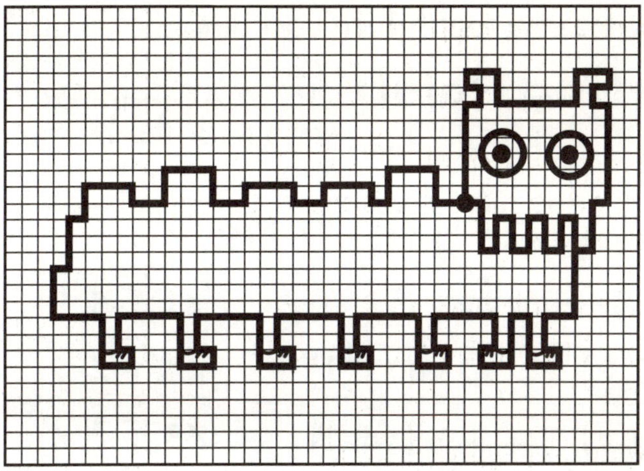

Paper: one-fourth-inch grid

Starting Point:

1. Place your paper with the long edge near the bottom of your desk.
2. Find the right end of the line closest to the top of your paper.
3. Count eleven full spaces to the left.
4. Move down twelve full spaces.
5. Make a pencil dot where the lines cross.
6. Put your pencil on the dot.

Move your pencil the following number of spaces:

1. Up 6
2. Right 1
3. Up 1
4. Left 1
5. Up 1
6. Right 2
7. Down 2
8. Right 5
9. Up 2
10. Right 2
11. Down 1
12. Left 1
13. Down 1
14. Right 1
15. Down 6
16. Left 1
17. Down 3
18. Left 1
19. Up 2
20. Left 1
21. Down 2
22. Left 1
23. Up 2
24. Left 1
25. Down 2
26. Left 1
27. Up 2
28. Left 1
29. Down 2
30. Left 1
31. Up 3
32. Left 3
33. Up 2
34. Left 3
35. Down 2
36. Left 1
37. Up 1
38. Left 3
39. Down 1
40. Left 2
41. Up 1
42. Left 3
43. Down 1
44. Left 2
45. Up 2
46. Left 3
47. Down 2
48. Left 2
49. Up 1
50. Left 3
51. Down 2
52. Left 1
53. Down 3
54. Left 1
55. Down 3
56. Right 3
57. Down 3
58. Right 2
59. Up 1
60. Left 1
61. Up 2
62. Right 4
63. Down 3
64. Right 2
65. Up 1
66. Left 1
67. Up 2
68. Right 4
69. Down 3
70. Right 2
71. Up 1
72. Left 1
73. Up 2
74. Right 4
75. Down 3
76. Right 2
77. Up 1
78. Left 1
79. Up 2
80. Right 4
81. Down 3
82. Right 2
83. Up 1
84. Left 1
85. Up 2
86. Right 4
87. Down 2
88. Left 1
89. Down 1
90. Right 2
91. Up 3
92. Right 1
93. Down 3
94. Right 2
95. Up 1
96. Left 1
97. Up 2
98. Right 2
99. Up 4

Add: eyes for the dragon and shoes for the dragon dancers

0-7682-0800-9 *Season's Griddings*

Chinese Dragon

Paper: one-fourth-inch coordinate axis grid

Starting Point: Place your paper with the long edge near the bottom of your desk. Start at (9, 2).

Move your pencil to:

1. (9, 8)
2. (10, 8)
3. (10, 9)
4. (9, 9)
5. (9, 10)
6. (11, 10)
7. (11, 8)
8. (16, 8)
9. (16, 10)
10. (18, 10)
11. (18, 9)
12. (17, 9)
13. (17, 8)
14. (18, 8)
15. (18, 2)
16. (17, 2)
17. (17, –1)
18. (16, –1)
19. (16, 1)
20. (15, 1)
21. (15, –1)
22. (14, –1)
23. (14, 1)
24. (13, 1)
25. (13, –1)
26. (12, –1)
27. (12, 1)
28. (11, 1)
29. (11, –1)
30. (10, –1)
31. (10, 2)
32. (7, 2)
33. (7, 4)
34. (4, 4)
35. (4, 2)
36. (3, 2)
37. (3, 3)
38. (0, 3)
39. (0, 2)
40. (–2, 2)
41. (–2, 3)
42. (–5, 3)
43. (–5, 2)
44. (–7, 2)
45. (–7, 4)
46. (–10, 4)
47. (–10, 2)
48. (–12, 2)
49. (–12, 3)
50. (–15, 3)
51. (–15, 1)
52. (–16, 1)
53. (–16, –2)
54. (–17, –2)
55. (–17, –5)
56. (–14, –5)
57. (–14, –8)
58. (–12, –8)
59. (–12, –7)
60. (–13, –7)
61. (–13, –5)
62. (–9, –5)
63. (–9, –8)
64. (–7, –8)
65. (–7, –7)
66. (–8, –7)
67. (–8, –5)
68. (–4, –5)
69. (–4, –8)
70. (–2, –8)
71. (–2, –7)
72. (–3, –7)
73. (–3, –5)
74. (1, –5)
75. (1, –8)
76. (3, –8)
77. (3, –7)
78. (2, –7)
79. (2, –5)
80. (6, –5)
81. (6, –8)
82. (8, –8)
83. (8, –7)
84. (7, –7)
85. (7, –5)
86. (11, –5)
87. (11, –7)
88. (10, –7)
89. (10, –8)
90. (12, –8)
91. (12, –5)
92. (13, –5)
93. (13, –8)
94. (15, –8)
95. (15, –7)
96. (14, –7)
97. (14, –5)
98. (16, –5)
99. (16, –1)

Add: eyes for the dragon and shoes for the dragon dancers

Vinnie's Valentine

Paper: one-half-inch grid

Starting Point:

1. Place your paper with the long edge near the bottom of your desk.
2. Find the left end of the line closest to the top of your paper.
3. Count one full space to the right.
4. Move down five full spaces.
5. Make a pencil dot where the lines cross.
6. Put your pencil on the dot.

Move your pencil the following number of spaces:

1. Right 3
2. Down 1
3. Right 1
4. Up 2
5. Right 1
6. Up 1
7. Right 1
8. Up 1
9. Right 2
10. Down 1
11. Right 1
12. Down 2
13. Right 1
14. Up 2
15. Right 1
16. Up 1
17. Right 2
18. Down 1
19. Right 1
20. Down 1
21. Right 1
22. Down 2
23. Right 1
24. Up 1
25. Right 1
26. Down 1
27. Right 1
28. Down 1
29. Left 1
30. Down 1
31. Left 1
32. Up 1
33. Left 1
34. Down 1
35. Left 1
36. Down 1
37. Left 1
38. Down 1
39. Left 1
40. Down 1
41. Left 1
42. Down 1
43. Left 1
44. Down 1
45. Left 1
46. Up 1
47. Left 1
48. Up 1
49. Left 1
50. Up 1
51. Left 1
52. Up 1
53. Left 1
54. Up 1
55. Left 1
56. Up 1
57. Left 1
58. Down 1
59. Left 3
60. Up 1
61. Right 1
62. Up 1
63. Left 1
64. Up 1

0-7682-0800-9 *Season's Griddings*

Crazy Cupid

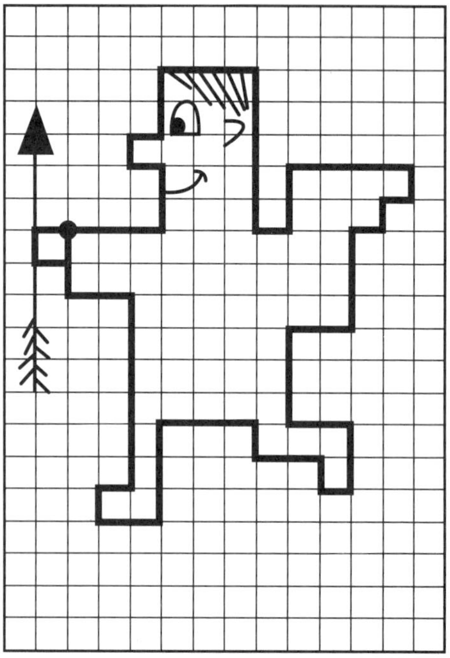

Paper: one-half-inch grid

Starting Point:

1. Place your paper with the short edge near the bottom of your desk.
2. Find the left end of the line closest to the top of your paper.
3. Count two full spaces to the right.
4. Move down seven full spaces.
5. Make a pencil dot where the lines cross.
6. Put your pencil on the dot.

Move your pencil:

1. Down 2 spaces
2. Right 2 spaces
3. Down 6 spaces
4. Left 1 space
5. Down 1 space
6. Right 2 spaces
7. Up 3 spaces
8. Right 3 spaces
9. Down 1 space
10. Right 2 spaces
11. Down 1 space
12. Right 1 space
13. Up 2 spaces
14. Left 2 spaces
15. Up 3 spaces
16. Right 2 spaces
17. Up 3 spaces
18. Right 1 space
19. Up 1 space
20. Right 1 space
21. Up 1 space
22. Left 4 spaces
23. Down 2 spaces
24. Left 1 space
25. Up 5 spaces
26. Left 3 spaces
27. Down 2 spaces
28. Left 1 space
29. Down 1 space
30. Right 1 space
31. Down 2 spaces
32. Left 4 spaces
33. Down 1 space
34. Right 1 space

Add: hair, ear, eye, mouth, and arrow

Vinnie's Valentine

Paper: one-half-inch letter coordinate grid

Starting Point: Place your paper with the long lettered edge near the bottom of your desk. Start at A–9.

Move your pencil to:

1. D–9
2. D–8
3. E–8
4. E–10
5. F–10
6. F–11
7. G–11
8. G–12
9. I–12
10. I–11
11. J–11
12. J–9
13. K–9
14. K–11
15. L–11
16. L–12
17. N–12
18. N–11
19. O–11
20. O–10
21. P–10
22. P–8
23. Q–8
24. Q–9
25. R–9
26. R–8
27. S–8
28. S–7
29. R–7
30. R–6
31. Q–6
32. Q–7
33. P–7
34. P–6
35. O–6
36. O–5
37. N–5
38. N–4
39. M–4
40. M–3
41. L–3
42. L–2
43. K–2
44. K–1
45. J–1
46. J–2
47. I–2
48. I–3
49. H–3
50. H–4
51. G–4
52. G–5
53. F–5
54. F–6
55. E–6
56. E–7
57. D–7
58. D–6
59. A–6
60. A–7
61. B–7
62. B–8
63. A–8
64. A–9

Crazy Cupid

Paper: one-half-inch letter coordinate grid

Starting Point: Place your paper with the long lettered edge near the bottom of your desk. Start at L–12.

Move your pencil to:

1. I–12
2. I–10
3. D–10
4. D–11
5. C–11
6. C–9
7. F–9
8. F–6
9. E–6
10. E–4
11. D–4
12. D–3
13. F–3
14. F–5
15. I–5
16. I–3
17. L–3
18. L–2
19. M–2
20. M–1
21. N–1
22. N–5
23. L–5
24. L–6
25. Q–6
26. Q–9
27. O–9
28. O–10
29. N–10
30. N–9
31. L–9
32. L–13
33. K–13
34. K–12

Turn your paper and add: hair, ear, eye, mouth and arrow

In Like a Lion!

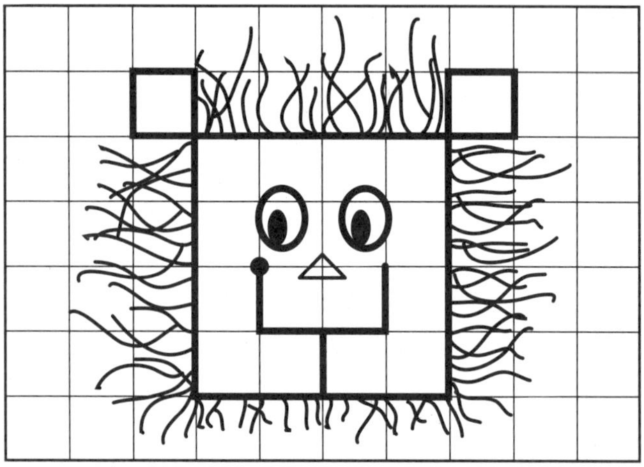

Paper: one-inch grid

Starting Point:

1. Place your paper with the long edge near the bottom of your desk.
2. Find the left end of the line closest to the bottom of your paper.
3. Count four full spaces to the right.
4. Move up three full spaces.
5. Make a pencil dot where the lines cross.
6. Put your pencil on the dot.

Move your pencil:

1. Down 1 space
2. Right 1 space
3. Down 1 space
4. Left 2 spaces
5. Up 4 spaces
6. Left 1 space
7. Up 1 space
8. Right 1 space
9. Down 1 space
10. Right 4 spaces
11. Up 1 space
12. Right 1 space
13. Down 1 space
14. Left 1 space
15. Down 4 spaces
16. Left 2 spaces
17. Up 1 space

    (return over your line)
18. Right 1 space
19. Up 1 space

Add: eyes, nose, and hair

Out Like a Lamb!

Paper: one-inch grid

Starting Point:

1. Place your paper with the long edge near the bottom of your desk.
2. Find the left end of the line closest to the top of your paper.
3. Count two full spaces to the right.
4. Move down one full space.
5. Make a pencil dot where the lines cross.
6. Put your pencil on the dot.

Move your pencil:

1. Up 1 space
2. Right 8 spaces
3. Down 7 spaces
4. Left 1 space
5. Up 2 spaces
6. Left 1 space
7. Down 2 spaces
8. Left 1 space
9. Up 2 spaces
10. Left 2 spaces
11. Down 2 spaces
12. Left 1 space
13. Up 2 spaces
14. Left 1 space
15. Down 2 spaces
16. Left 1 space
17. Up 3 spaces
18. Left 1 space
19. Up 2 spaces
20. Left 1 space
21. Up 1 space
22. Right 4 spaces
23. Down 1 space
24. Left 1 space
25. Down 2 spaces
26. Left 1 space

Add: eyes, nose, mouth, and curly wool

O'Leary Leprechaun

Paper: one-half-inch grid

Starting Point:

1. Place your paper with the short edge near the bottom of your desk.
2. Find the left end of the line closest to the top of your paper.
3. Count one full space to the right.
4. Move down nine full spaces.
5. Make a pencil dot where the lines cross.
6. Put your pencil on the dot.

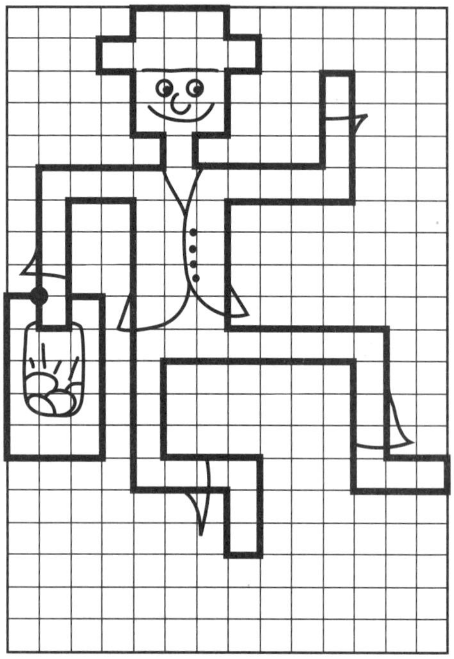

Move your pencil:

1. Down 1 space
2. Right 1 space
3. Up 1 space
4. Right 1 space
5. Down 5 spaces
6. Left 3 spaces
7. Up 5 spaces
8. Right 1 space
9. Up 4 spaces
10. Right 4 spaces
11. Up 1 space
12. Left 1 space
13. Up 2 spaces
14. Left 1 space
15. Up 1 space
16. Right 1 space
17. Up 1 space
18. Right 3 spaces
19. Down 1 space
20. Right 1 space
21. Down 1 space
22. Left 1 space
23. Down 2 spaces
24. Left 1 space
25. Down 1 space
26. Right 4 spaces
27. Up 3 spaces
28. Right 1 space
29. Down 4 spaces
30. Left 4 spaces
31. Down 4 spaces
32. Right 5 spaces
33. Down 4 spaces
34. Right 2 spaces
35. Down 1 space
36. Left 3 spaces
37. Up 4 spaces
38. Left 6 spaces
39. Down 3 spaces
40. Right 3 spaces
41. Down 3 spaces
42. Left 1 space
43. Up 2 spaces
44. Left 3 spaces
45. Up 9 spaces
46. Left 2 spaces
47. Down 3 spaces

Add: hat, eyes, nose, mouth, jacket, pants, and gold

April Fool!

Paper: one-fourth-inch grid

Starting Point:

1. Place your paper with the short edge near the bottom of your desk.
2. Find the left end of the line closest to the top of your paper.
3. Count seven full spaces to the right.
4. Move down thirteen full spaces.
5. Make a pencil dot where the lines cross.
6. Put your pencil on the dot.

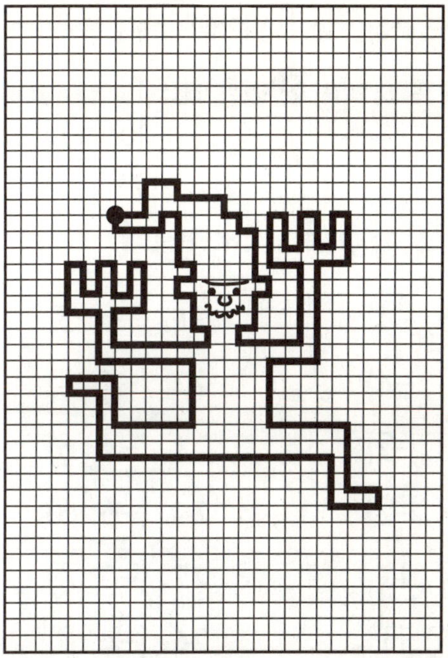

Move your pencil the following number of spaces:

1. Right 2
2. Up 2
3. Right 2
4. Down 1
5. Right 3
6. Down 1
7. Right 1
8. Down 1
9. Right 1
10. Down 3
11. Right 1
12. Down 1
13. Left 1
14. Down 2
15. Left 1
16. Down 1
17. Right 4
18. Up 5
19. Left 2
20. Up 3
21. Right 1
22. Down 2
23. Right 1
24. Up 2
25. Right 1
26. Down 2
27. Right 1
28. Up 2
29. Right 1
30. Down 3
31. Left 2
32. Down 6
33. Left 3
34. Down 4
35. Right 5
36. Down 4
37. Right 2
38. Down 1
39. Left 3
40. Up 3
41. Left 15
42. Up 4
43. Left 2
44. Up 1
45. Right 3
46. Down 3
47. Right 5
48. Up 4
49. Left 6
50. Up 3
51. Left 2
52. Up 3
53. Right 1
54. Down 2
55. Right 1
56. Up 2
57. Right 1
58. Down 2
59. Right 1
60. Up 2
61. Right 1
62. Down 3
63. Left 2
64. Down 2
65. Right 6
66. Up 1
67. Left 1
68. Up 2
69. Left 1
70. Up 1
71. Right 1
72. Up 1
73. Left 1
74. Up 3
75. Left 1
76. Down 1
77. Left 3
78. Up 1

Add: hat, eyes, nose, and mouth

April Fool!

Paper: one-fourth-inch coordinate axis grid

Starting Point: Place your paper with the long edge near the bottom of your desk. Start at (–6, –6).

Move your pencil to:

1. (–6, –4)
2. (–8, –4)
3. (–8, –2)
4. (–7, –2)
5. (–7, 1)
6. (–6, 1)
7. (–6, 2)
8. (–5, 2)
9. (–5, 3)
10. (–2, 3)
11. (–2, 4)
12. (–1, 4)
13. (–1, 3)
14. (1, 3)
15. (1, 2)
16. (2, 2)
17. (2, 6)
18. (–3, 6)
19. (–3, 4)
20. (–6, 4)
21. (–6, 5)
22. (–4, 5)
23. (–4, 6)
24. (–6, 6)
25. (–6, 7)
26. (–4, 7)
27. (–4, 8)
28. (–6, 8)
29. (–6, 9)
30. (–3, 9)
31. (–3, 7)
32. (3, 7)
33. (3, 4)
34. (7, 4)
35. (7, 9)
36. (11, 9)
37. (11, 11)
38. (12, 11)
39. (12, 8)
40. (9, 8)
41. (9, –7)
42. (5, –7)
43. (5, –9)
44. (4, –9)
45. (4, –6)
46. (7, –6)
47. (7, –1)
48. (3, –1)
49. (3, –7)
50. (0, –7)
51. (0, –9)
52. (–3, –9)
53. (–3, –8)
54. (–1, –8)
55. (–1, –7)
56. (–3, –7)
57. (–3, –6)
58. (–1, –6)
59. (–1, –5)
60. (–3, –5)
61. (–3, –4)
62. (0, –4)
63. (0, –6)
64. (2, –6)
65. (2, 0)
66. (1, 0)
67. (1, –1)
68. (–1, –1)
69. (–1, –2)
70. (–2, –2)
71. (–2, –1)
72. (–3, –1)
73. (–3, –2)
74. (–6, –2)
75. (–6, –3)
76. (–5, –3)
77. (–5, –6)
78. (–6, –6)

Turn your paper and add: hat, eyes, nose, and mouth

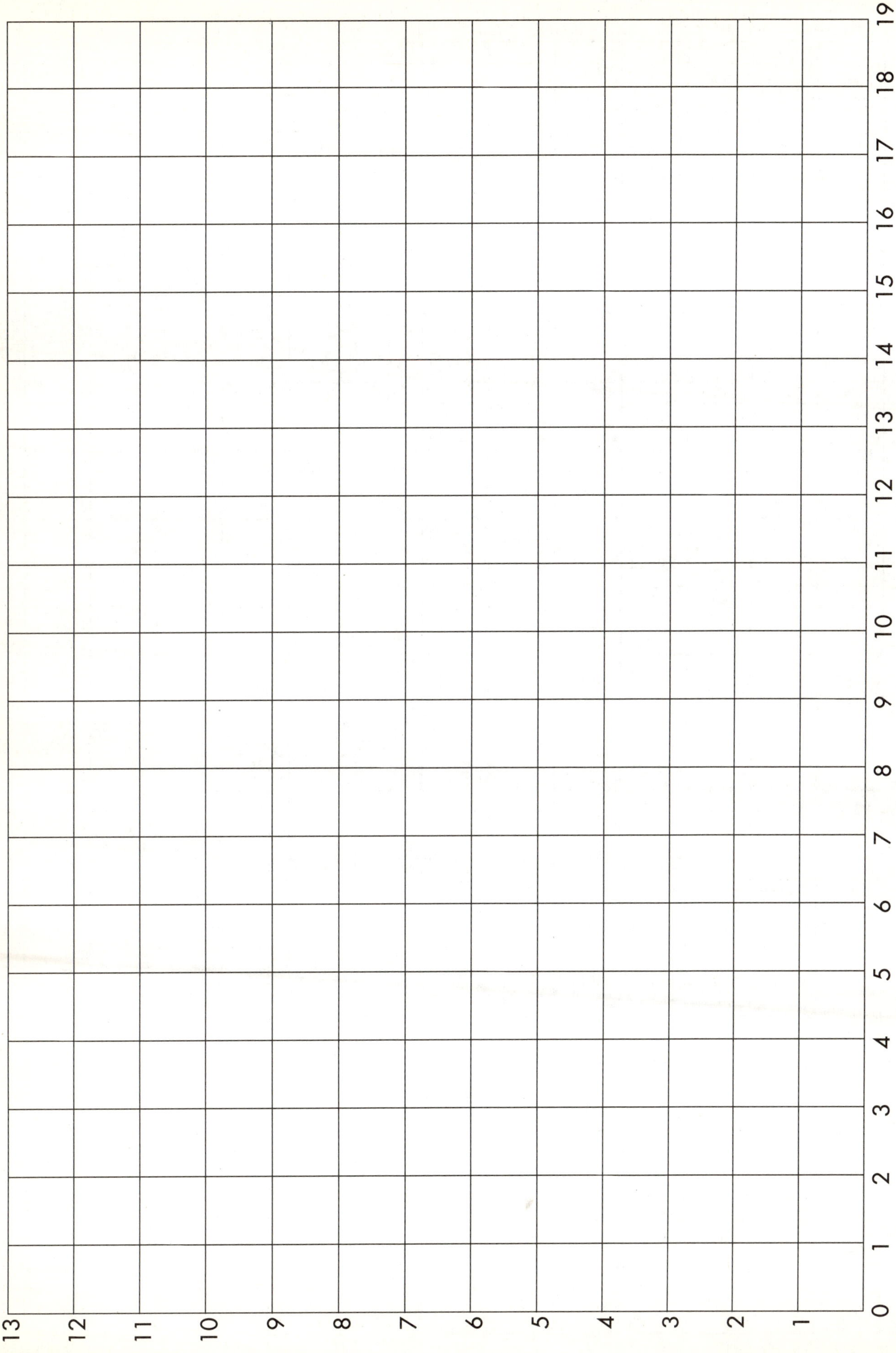
0 1 2 3 4 5 6 7 8 9 10 11 12 13 14 15 16 17 18 19
1 2 3 4 5 6 7 8 9 10 11 12 13

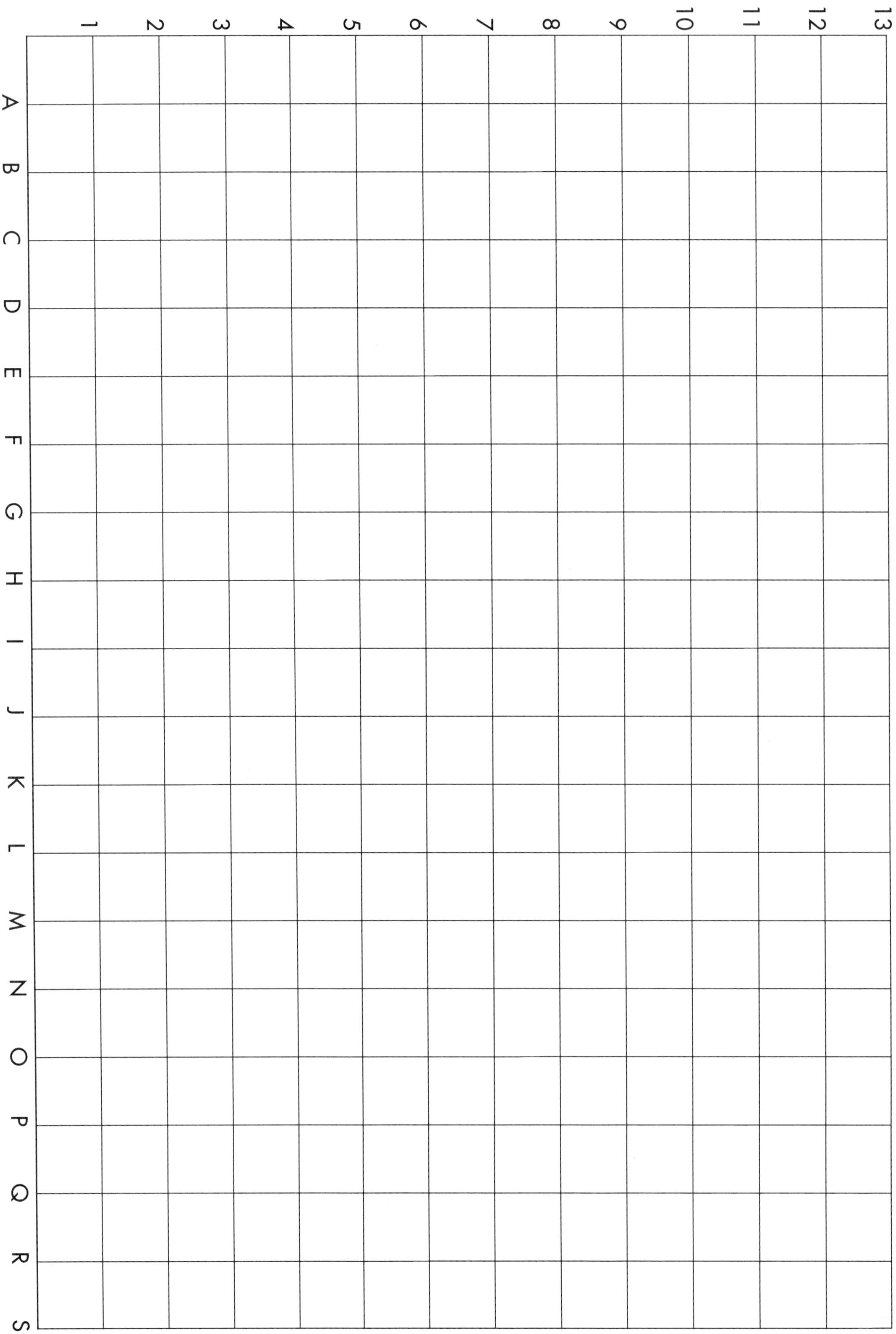
1 2 3 4 5 6 7 8 9 10 11 12 13
A B C D E F G H I J K L M N O P Q R S

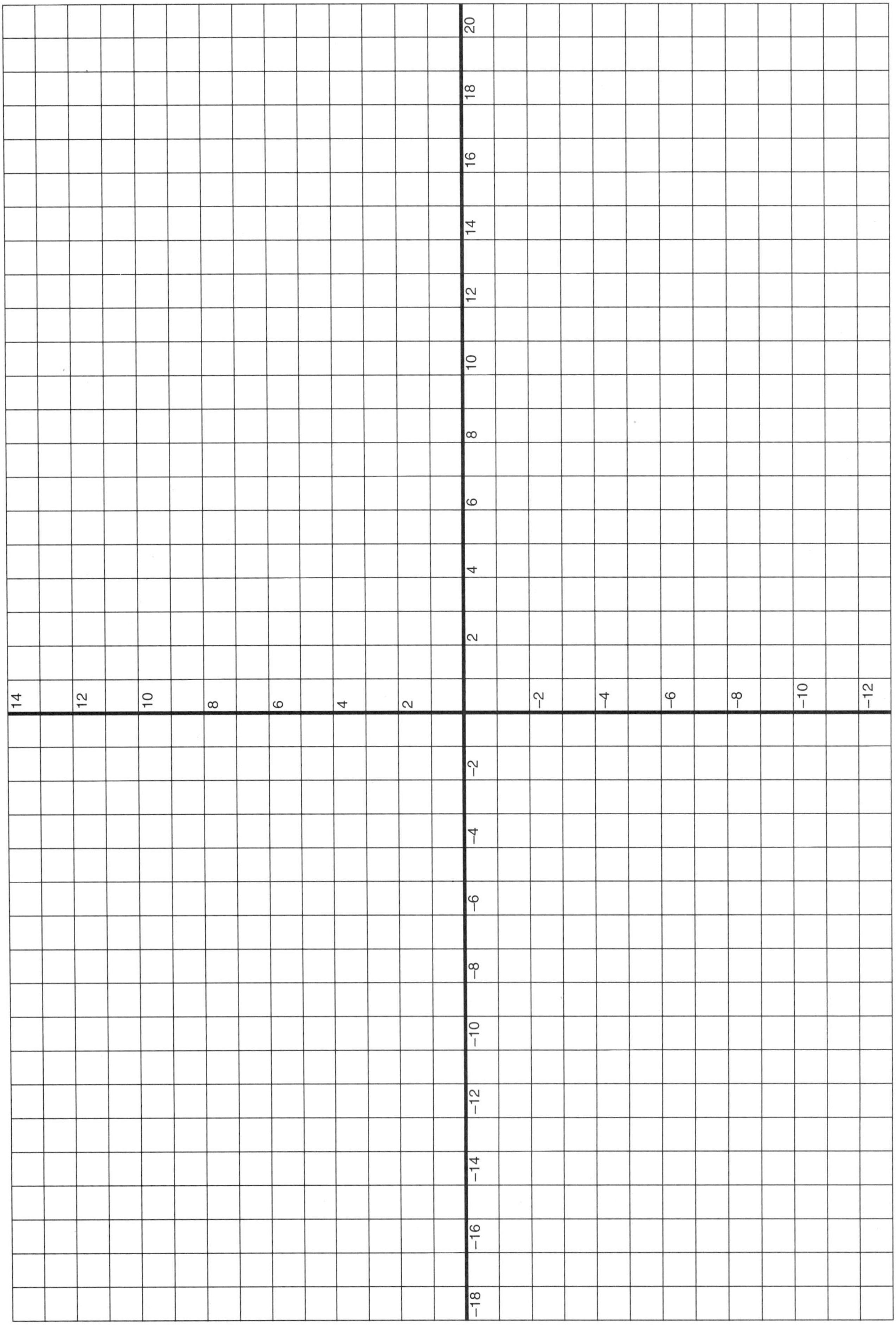
20
18
16
14
12
10
8
6
4
2
–2
–4
–6
–8
–10
–12
–14
–16
–18
14
12
10
8
6
4
2
–2
–4
–6
–8
–10
–12